100
things you should know about
FLIGHT

100

things you should know about

FLIGHT

Sue Becklake

Consultant: Chris Oxlade

MC
PUBLISHERS

This 2009 edition published and distributed by:
Mason Crest Publishers Inc.
370 Reed Road, Broomall, Pennsylvania 19008
(866) MCP-BOOK (toll free)
www.masoncrest.com

Library of Congress Cataloging-in-Publication data is available
100 Things You Should Know About Flight
ISBN 978-1-4222-1521-0

100 Things You Should Know About - 10 Title Series
ISBN 978-1-4222-1517-3

Printed in the United States of America

First published in 2008 by Miles Kelly Publishing Ltd
Bardfield Centre, Great Bardfield, Essex, CM7 4SL

ACKNOWLEDGEMENTS
The publishers would like to thank the following artists
who have contributed to this book:
Mike Foster, Terry Pastor, Richard Roussel, Tim@kja-artists.com
Cover artwork by Mike Saunders

All other artworks are from the Miles Kelly Artwork Bank

The publishers would like to thank the following sources
for the use of their photographs:
Page 8 Roger-Viollet/TopFoto; 13 sharply_done-Fotolia.com; 14 Bettmand/Corbis;
15(b) Chris Fourie-Fotolia.com; 21(t) Ron Watts/Corbis, (b) Antony Nettle/Alamy;
22 Arne Dedart/dpa/Corbis; 29(t) Michel Setboun/Corbis, (b) Jetpics-Fotolia.com;
30(b) IL-Fotolia.com; 31(t) Star Images/Topfoto, (b) Lowell Georgia/Corbis;
33 Corbis; 35(b) TopFoto/ImageWorks; 36(t) George Hall/Corbis, (b) George Hall/Corbis;
37 Charles O'Rear/Corbis; 42 TopFoto.co.uk; 46(b) TopFoto/keystone; 47(b) NASA

All other photographs are from:
Castrol, Corel, digitalSTOCK, digitalvision, John Foxx, PhotoAlto,
PhotoDisc, PhotoEssentials, PhotoPro, Stockbyte

Contents

Flying machines

Modern fighter planes such as these Eurofighter Typhoons speed through the skies to attack an enemy.

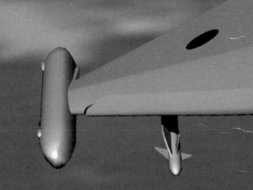

1 **People first tried to fly hundreds of years ago.** They made wings from different materials and invented machines to take to the air, but nothing seemed to work. Then it was discovered that a curved wing moving through the air lifted upwards – and the age of flight began. Now huge airliners carry hundreds of passengers halfway around the world, high above the clouds every day – and in the future, passengers may be able to travel into space.

The first flights

2 **The first humans to fly went up in a balloon, not a plane.** In 1783 in France, the Mongolfier brothers made a large balloon and heated the air inside it by lighting a fire underneath. The balloon rose into the air carrying two passengers who floated over Paris for 25 minutes. They carried a small fire to heat the air and keep the balloon afloat.

In 1783 two volunteers flew to a height of half a mile in the Montgolfier brothers' balloon.

3 **The first planes were gliders with no engines.** In the 1890s, a German called Otto Lilienthal experimented with flying by building small gliders that would carry him into the air. He tried using different wing shapes to see which worked best. To become airborne, he launched himself from the top of a hill and glided downwards.

I DON'T BELIEVE IT!

Before letting people fly in their balloon, the Montgolfier brothers tried it out with a duck, a sheep and a chicken!

To test his flying machines Otto Lilienthal practised jumping from cliffs, hills, and rooftops.

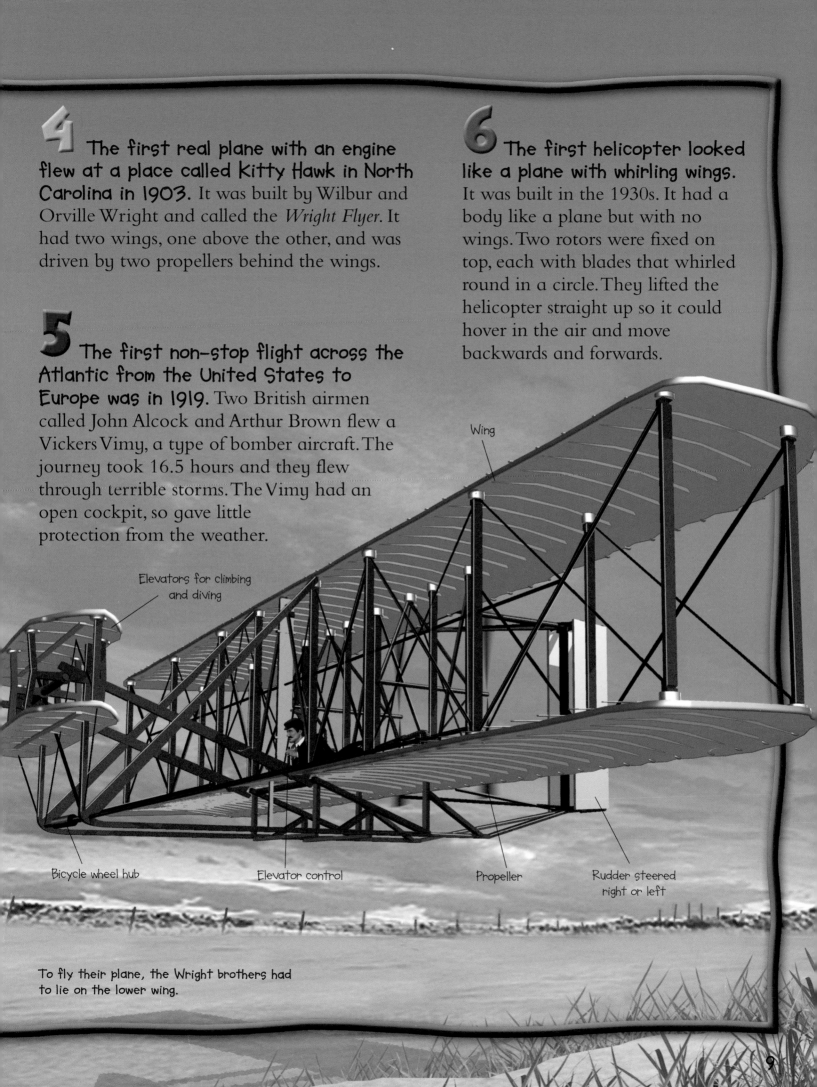

4 **The first real plane with an engine flew at a place called Kitty Hawk in North Carolina in 1903.** It was built by Wilbur and Orville Wright and called the *Wright Flyer*. It had two wings, one above the other, and was driven by two propellers behind the wings.

5 **The first non-stop flight across the Atlantic from the United States to Europe was in 1919.** Two British airmen called John Alcock and Arthur Brown flew a Vickers Vimy, a type of bomber aircraft. The journey took 16.5 hours and they flew through terrible storms. The Vimy had an open cockpit, so gave little protection from the weather.

6 **The first helicopter looked like a plane with whirling wings.** It was built in the 1930s. It had a body like a plane but with no wings. Two rotors were fixed on top, each with blades that whirled round in a circle. They lifted the helicopter straight up so it could hover in the air and move backwards and forwards.

Wing

Elevators for climbing and diving

Bicycle wheel hub

Elevator control

Propeller

Rudder steered right or left

To fly their plane, the Wright brothers had to lie on the lower wing.

Early days of flying

7 During World War 1, planes were used in battle. Many were biplanes with double wings such as the Sopwith Camel. Fighter planes were built with machine guns to shoot down enemy planes. They could also fire at soldiers on the ground, while larger planes and airships dropped bombs.

A British pilot in his Sopwith Camel watches a burning German plane. The plane got its name from the hump over its guns.

The *Hindenburg*

8 In the 1920s and '30s huge airships carried people between Europe and the United States. These giant oval machines had engines and propellers to push and steer them along. Passengers travelled in cabins below the airship. However the light gas used to lift the airship was often hydrogen, which easily catches fire. In May 1937, the airship *Hindenburg* burst into flames. This put an end to travel by airship.

Thirty-six people died when the *Hindenburg* airship crashed in flames in 1937.

TRUE OR FALSE?

1. A fight between two planes was called a catfight.
2. Jet airliners can fly higher than propeller aircraft.
3. Airships are filled with heavy gas.

Answers:
1. False 2. True 3. False

9 Planes played an essential part during World War II. Large bombers could carry a heavy load of bombs to drop on factories, ports and cities. However they were not fast enough to escape attack by enemy fighter planes. The fighters were built for speedy diving and turning to attack enemy planes. Duels between fighter planes in the air were called dogfights.

The first jet fighter, the German Messerschmitt 262, flew in 1944. It carried out bombing raids, but its heavy bombs slowed it down.

10 Flying boats were planes that could land on water. Shaped like a boat underneath, they could fly to places where there were no runways. They took passengers and mail all round the world in the 1930s and '40s. The largest flying boats had room for about 70 passengers.

The Comet airliner had four jet engines, two in each wing.

11 After World War II, more people began to travel by plane. New jet engines allowed planes to fly faster and higher and carry more passengers. The first jet airliner began carrying passengers in 1952. It was called the Comet and could fly twice as fast as propeller-driven planes. Big, modern airliners with jet engines now carry hundreds of passengers on long journeys every day.

Parts of a plane

12 The main body of a plane is usually long and thin, with a pointed nose and smooth shape to cut through the air easily. This is called the fuselage. At the front is the cockpit, or flight deck, where the pilot controls the plane. In a passenger plane, most of the remaining body is taken up by a cabin with seats for passengers. Under the the cabin floor is a hold to store luggage.

An Airbus A380 passenger liner. A typical airliner such as this has a smooth, streamlined body.

Tail

Fuselage

Engine

Cockpit (flight deck)

13 The wings keep the plane up in the air and stick out on either side of the fuselage. They are long and thin with a curved top surface. Engines are usually attached to the wings. Fuel tanks are inside the wings. Along the front and back edges of the wings are moving parts called control surfaces, which can be tilted up or down to steer the plane.

I DON'T BELIEVE IT!

A 747 Jumbo Jet has 18 wheels in total – a set of two wheels under the nose and four sets of four wheels under the body and wings.

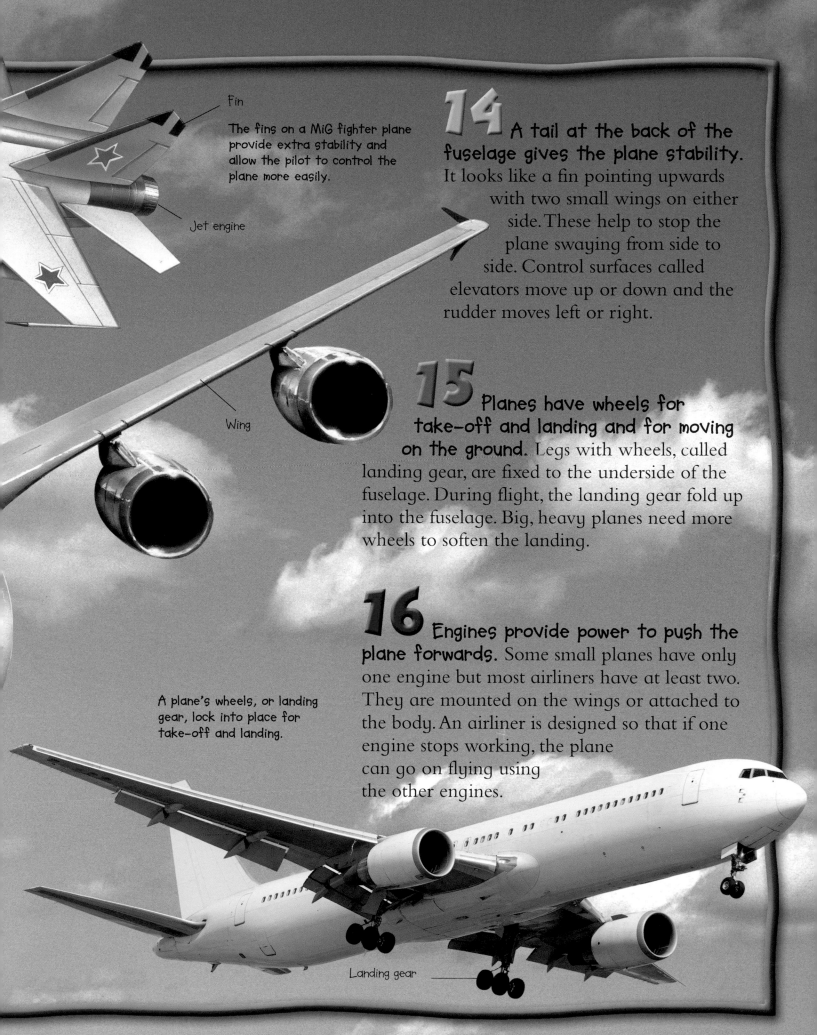

Fin

The fins on a MiG fighter plane provide extra stability and allow the pilot to control the plane more easily.

Jet engine

Wing

14 A tail at the back of the fuselage gives the plane stability. It looks like a fin pointing upwards with two small wings on either side. These help to stop the plane swaying from side to side. Control surfaces called elevators move up or down and the rudder moves left or right.

15 Planes have wheels for take-off and landing and for moving on the ground. Legs with wheels, called landing gear, are fixed to the underside of the fuselage. During flight, the landing gear fold up into the fuselage. Big, heavy planes need more wheels to soften the landing.

A plane's wheels, or landing gear, lock into place for take-off and landing.

16 Engines provide power to push the plane forwards. Some small planes have only one engine but most airliners have at least two. They are mounted on the wings or attached to the body. An airliner is designed so that if one engine stops working, the plane can go on flying using the other engines.

Landing gear

How planes fly

17 **A plane flies by moving through the air.** The engines drive the plane forwards with a force called thrust. However, air pushes in the opposite direction and slows the plane down. This is called drag. Weight is the force that tries to pull the plane down. When moving through the air, the wings give an upward force called lift.

Direction of air flow around wing

As the wing moves forward, air streams under and over it, lifting it up.

18 **Air flowing over the wings gives an upward lift.** The wings are a special shape called an aerofoil. The top curves upwards while the bottom is flatter. As the plane moves forward, air flowing over the top has further to go and is more spread out than the air beneath. The air beneath pushes the wing harder than the air above it, so the wing lifts, taking the plane with it.

LIFTING FORCE

Wrap a strip of narrow paper around a pencil. Holding one end of the strip, blow hard over the top of it. Watch the free end of the paper lift upwards. This shows how an aircraft wing lifts as it moves through the air, keeping the heavy aircraft in the air. The faster you blow, the higher the paper lifts.

A force called lift pulls the plane up

A force called thrust pulls the plane forward

A force called weight pulls the plane down

A flying plane is pushed and pulled by four different forces in four different directions.

Flaps on the wings, called ailerons, direct the air flow up or down.

Aileron

20
As the plane moves forwards it pushes against the air. The air pushes back, which slows the plane down and makes it use more fuel. Aircraft builders try to make the drag as minimal as possible by designing the plane to be smooth and streamlined so it cuts cleanly through the air.

19
The engines give the thrust that drives the plane forwards in the air. As the plane travels faster, the lifting force grows stronger. This force must be equal to the weight of the plane before it can rise into the air and fly. This means that the thrust from the engines must drive the plane quickly to give it enough lift to fly.

21
The weight of a plane is always trying to pull it down. For this reason, planes are built to be as light as possible, using light but strong materials. Even so, a Boeing 747 jumbo jet with all its passengers and luggage can weigh as much as 360 tons and still take off.

Jet engines or propellers thrust a plane forwards.

Propeller engine

A force called drag pulls the plane back

22
Planes get thrust from jet engines or propellers. Jet engines are more powerful and better for flying high up where the air is thinner. Airliners and fighter planes have jet engines. Propellers are more useful for planes that fly slower and nearer the ground. Most small private planes and some large planes that carry heavy cargo use propeller engines.

Jet engine

Powerful engines

23 **A jet engine thrusts a plane forwards by shooting out a jet of hot gases.** A turbojet engine uses spinning blades called a compressor to suck air into the front of the engine and squeeze it tightly. This air is then mixed with fuel inside the engine, as the fuel requires air to burn. The burning fuel creates hot gases that shoot out of a nozzle at the back of the engine.

BALLOON JET

Blow up a balloon then let it go. Watch the balloon shoot away as the air rushes out. In the same way, a plane shoots forward when gases rush out of its jet engines.

Fuel is mixed with air and then burnt

Compressor sucks in air

In a turbojet engine, air is sucked in and burnt with fuel to create hot gases.

Exhaust gases

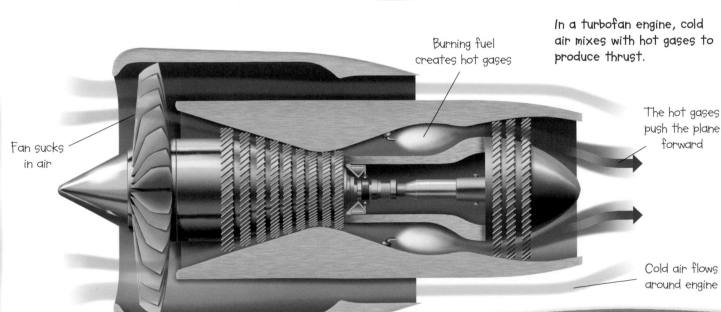

Burning fuel creates hot gases

In a turbofan engine, cold air mixes with hot gases to produce thrust.

Fan sucks in air

The hot gases push the plane forward

Cold air flows around engine

24
A turbofan engine is another type of jet engine used by modern airliners. These are less noisy than turbojet engines and cheaper to run. A large fan at the front sucks in air, but not all of it is squeezed and mixed with fuel. Some of the air flows around the outside of the engine and mixes with the hot gases shooting out of the back.

25
Propellers whiz round at high speed, pulling the plane through the air. The propeller has two or more blades sticking out from the centre. Each blade is like a small wing and as it spins, it pushes the air backwards so the plane moves forwards. Small planes have just one propeller at the front, but larger planes may have two or more propellers, each driven by its own engine.

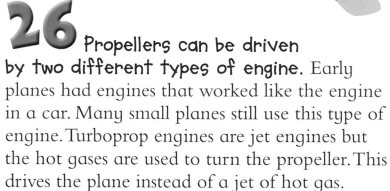

Propeller blade

Hub

When the propellers spin, they pull the plane through the air.

26
Propellers can be driven by two different types of engine. Early planes had engines that worked like the engine in a car. Many small planes still use this type of engine. Turboprop engines are jet engines but the hot gases are used to turn the propeller. This drives the plane instead of a jet of hot gas.

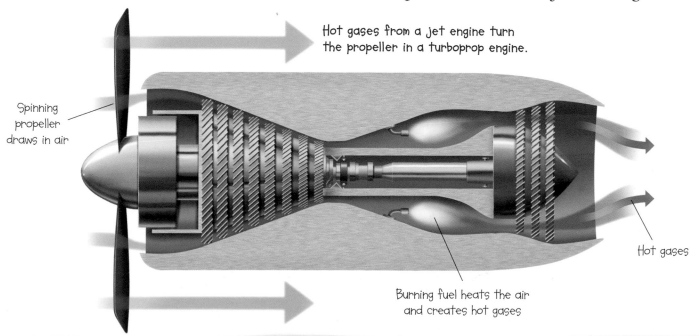

Hot gases from a jet engine turn the propeller in a turboprop engine.

Spinning propeller draws in air

Hot gases

Burning fuel heats the air and creates hot gases

Climbing, diving and turning

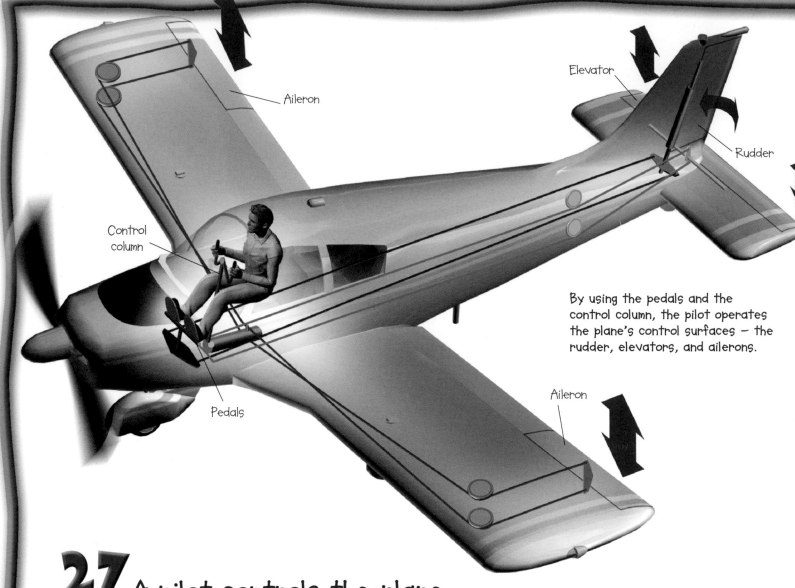

Aileron

Elevator

Rudder

Control column

Pedals

By using the pedals and the control column, the pilot operates the plane's control surfaces — the rudder, elevators, and ailerons.

Aileron

27 A pilot controls the plane, making it climb, dive or turn. He uses foot pedals and the control column. These are connected to the control surfaces on the wings and tail, which steer the plane. A plane moves in three directions. "Yaw" means to turn to the right or left. "Pitch" tilts the nose up or down and "roll" is simply to roll from side to side.

28 To make the plane climb higher the pilot pulls the control column towards him. This makes the elevators, the flaps attached to the back of the tail, tilt upwards. The air flowing over the elevators now pushes the tail down and so the nose goes up and the plane climbs up at an angle. To dive, the pilot pushes the control column forward, tilting the elevators down.

29 **Moving the control column to the left or right moves flaps called ailerons on the wings.** Pushing to the left makes the left aileron go up and the one on the right go down. This lowers the left wing and lifts the right wing so the plane rolls over to the left. Pushing the control column to the right rolls the plane over to the right.

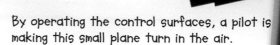

By operating the control surfaces, a pilot is making this small plane turn in the air.

30 **The pedals turn the plane to the right or left.** These are connected to the rudder on the fin. Pushing the left pedal swings the rudder to the left. This turns the tail to the right and the nose towards the left, so the plane makes a left turn. The right pedal swings the rudder to the right, and the plane makes a right turn.

Roll
Wing tilts up or down

Once in the air, a plane can move in three ways — roll, pitch, and yaw.

Pitch
Nose tilts up or down

Yaw
Turns left or right

31 **When the plane turns in the air it must roll at the same time.** This is called banking and is similar to a bike tilting over when turning a corner. For a right turn, the control column is moved to the right while the right pedal is pushed, turning and rolling the plane to the right at the same time.

Taking off and landing

32 **Planes take off by speeding along a runway.** The pilot sets the engines for maximum power. She travels down the runway until the plane has enough speed for the wings to lift it and fly. Planes usually take off facing into the wind. This gives a faster air flow over the wings and more lift.

After take-off, a plane climbs steeply until it reaches its cruising height.

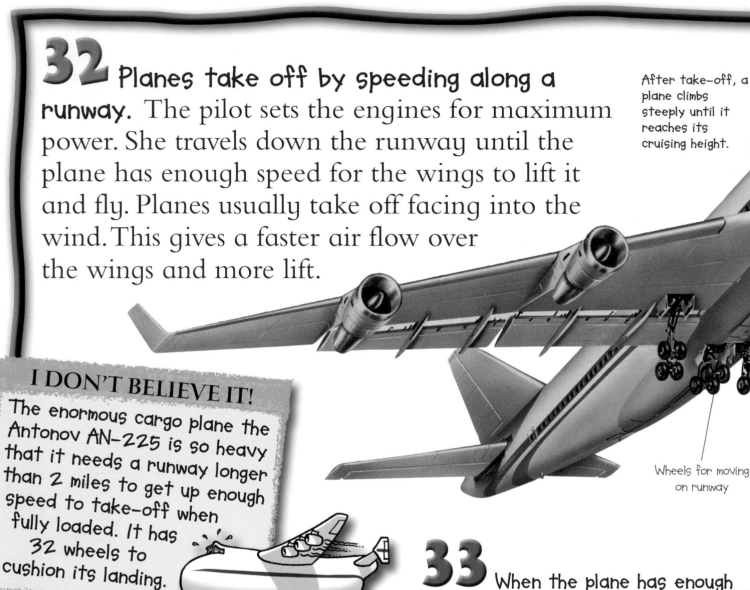

Wheels for moving on runway

I DON'T BELIEVE IT!

The enormous cargo plane the Antonov AN-225 is so heavy that it needs a runway longer than 2 miles to get up enough speed to take-off when fully loaded. It has 32 wheels to cushion its landing.

33 **When the plane has enough speed, the pilot pulls the control column back, raising the elevators on the tail.** This lifts the nose and the plane starts to climb into the air. As soon as the plane is climbing steadily the pilot folds the wheels up into the body. This reduces drag and allows the plane to speed up more quickly. It climbs to its cruising height then levels off for the journey. Airliners fly above the clouds where the air is thinner and there is less drag.

This plane has just taken off from an airport close to a holiday resort. Take-off is an impressive sight – but it can be very noisy.

34 Air traffic controllers give the pilot instructions and information so that the plane can take off and land safely. The pilot must always follow the instructions of the air traffic controllers and she cannot land or take-off until she has their permission to do so.

In the control tower, computers show air traffic controllers which planes are ready for take-off and landing.

35 Near the end of the journey, the pilot pushes the control column forward to point the nose down and descend. The pilot lines up the plane along the runway. She slows the engines and lowers the flaps on the wings to help slow the plane. When the wheels touch the ground, the plane may reverse its engines to stop.

36 In fog, when the pilot cannot see very far, planes can land automatically. The plane picks up radio signals from beacons beside the runway. The plane's computers use these signals to line the plane up with the runway and land safely on the runway.

When landing, the wheels under the plane's fuselage touch down on the runway first, followed by the nose wheels.

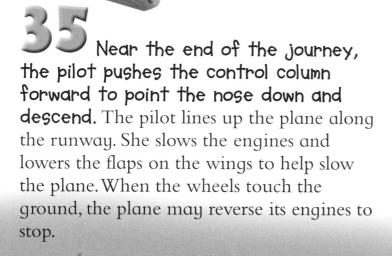

The flight deck

37 **The cockpit is where the pilot sits when he's flying the plane.** In larger planes this is called the flight deck. All around are displays, switches, and lights. In front of his seat is the control column and pedals for steering the plane.

A pilot carefully checks the controls on the flight deck of an Airbus A380 before take-off.

38 **The cockpit instruments tell the pilot all he needs to know about the plane.** There are displays showing speed, altitude (height), and whether the plane is climbing, diving, or rolling. There is also information about the weather. Warning lights alert the pilot of any problems. Older planes show all this information on dials but many modern planes use computer screens.

39 An automatic pilot can take over from the human pilot and fly the plane. As well as landing a plane automatically, an automatic pilot can fly the plane for much of the journey. The pilot sets the speed, height, and direction, and a computer controls the plane, making any necessary adjustments to keep it on course.

40 In larger planes, such as passenger airliners and cargo planes, the flight deck has two seats — one for the pilot and the other for the co-pilot. Each seat has its own set of controls with a control column and pedals so the pilot and co-pilot can take over from each other at any time.

41 Fighter planes often have only a single seat for the pilot. Modern fighters usually have a "glass" cockpit, with the displays on computer screens. Some of these will be "head-up" displays. This means that the information is shown on a glass screen in front of the pilot's eyes so he doesn't have to look down to see it. He can look out through the screen at the same time to see where he is going. Some head-up displays appear on a special helmet worn by the pilot.

A passenger jet airliner

42 In a passenger jet airliner, the cabin takes up almost all of the body. It has seats for the passengers, usually between 200 and 400 altogether. At high altitude, the air outside is too thin for people to breathe. This means that the cabin and cockpit are sealed and filled with air for the passengers and crew to breathe.

A Boeing 747 jumbo jet has wide wings that provide enormous lift. Its four powerful engines push the plane forward at cruising speeds of 620 miles per hour.

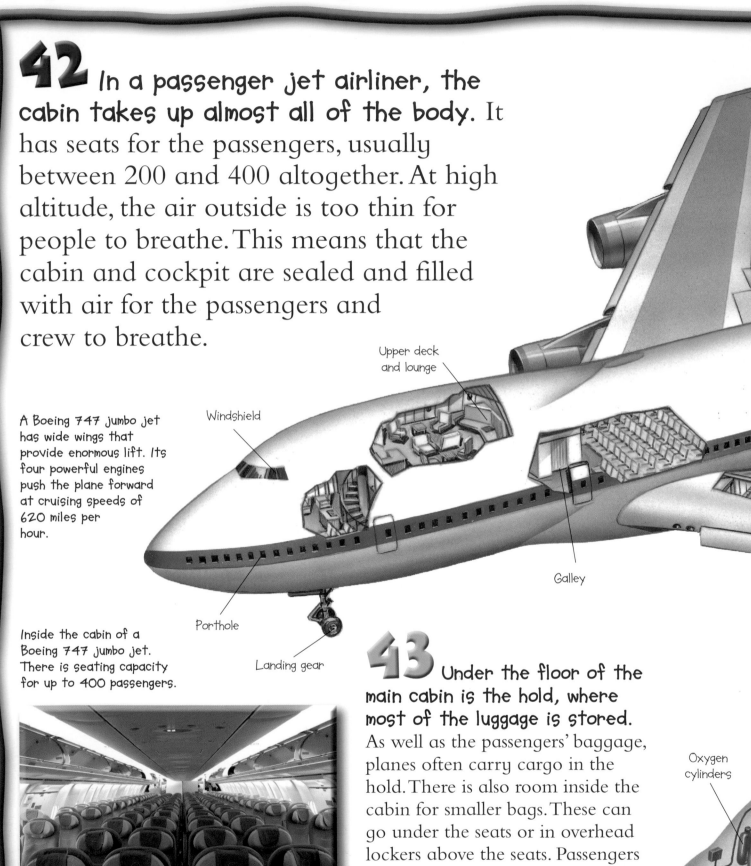

Upper deck and lounge

Windshield

Galley

Porthole

Landing gear

Oxygen cylinders

Inside the cabin of a Boeing 747 jumbo jet. There is seating capacity for up to 400 passengers.

43 Under the floor of the main cabin is the hold, where most of the luggage is stored. As well as the passengers' baggage, planes often carry cargo in the hold. There is also room inside the cabin for smaller bags. These can go under the seats or in overhead lockers above the seats. Passengers cannot take a lot of very heavy luggage because the plane cannot fly if it is too heavy.

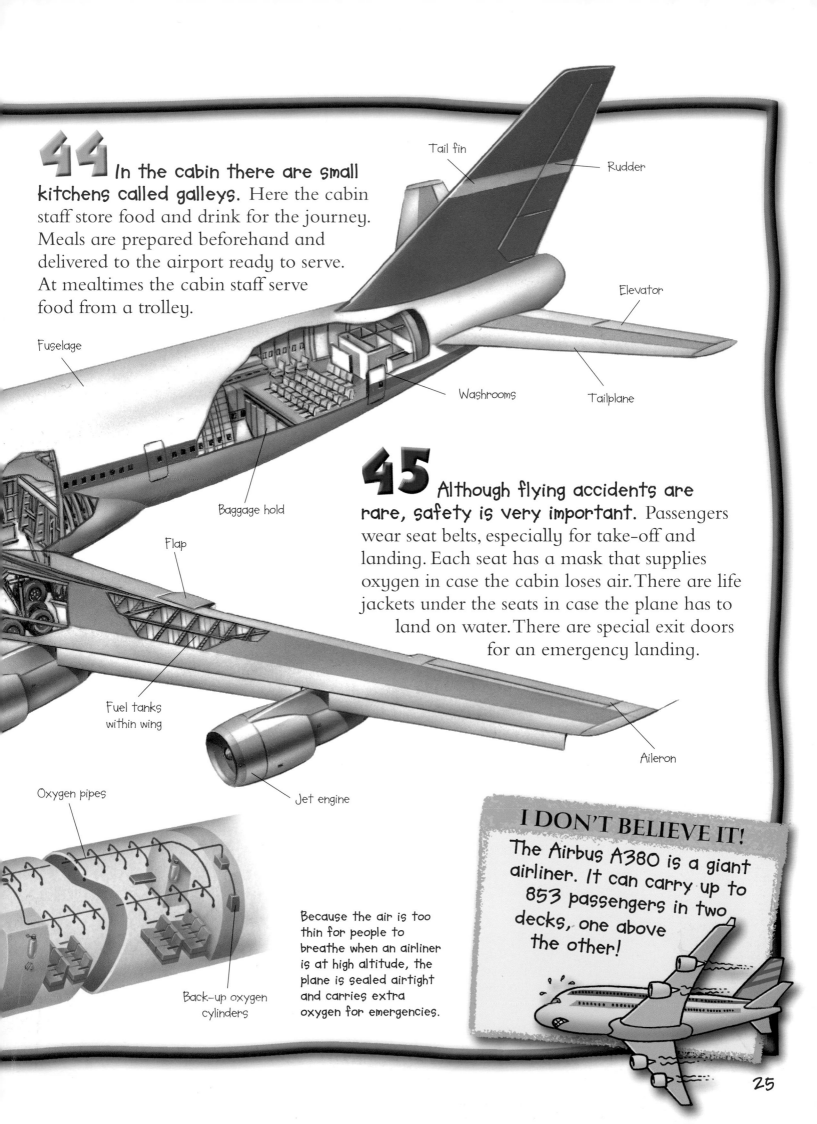

44 In the cabin there are small kitchens called galleys. Here the cabin staff store food and drink for the journey. Meals are prepared beforehand and delivered to the airport ready to serve. At mealtimes the cabin staff serve food from a trolley.

Tail fin

Rudder

Fuselage

Elevator

Washrooms

Tailplane

Baggage hold

Flap

45 Although flying accidents are rare, safety is very important. Passengers wear seat belts, especially for take-off and landing. Each seat has a mask that supplies oxygen in case the cabin loses air. There are life jackets under the seats in case the plane has to land on water. There are special exit doors for an emergency landing.

Fuel tanks within wing

Aileron

Oxygen pipes

Jet engine

Back-up oxygen cylinders

Because the air is too thin for people to breathe when an airliner is at high altitude, the plane is sealed airtight and carries extra oxygen for emergencies.

I DON'T BELIEVE IT!
The Airbus A380 is a giant airliner. It can carry up to 853 passengers in two decks, one above the other!

At the airport

46 The most obvious parts of an airport are the terminals and the runways. Most big airports have at least two runways over a mile and a half long. These allow the largest jets to take off and land. Lights and markings along the runway show the pilot where to touch down.

47 Passengers arrive at the terminal building for their flights. Tickets and passports are checked and luggage is left with the airline staff. Then passengers and their hand luggage are checked by security officers for dangerous objects such as knives. The luggage is also checked. People wait to board their plane in the departure lounge.

Passenger boarding bridge

Airports can be huge and stretch for several miles. They have a constant flow of planes that are taking off and landing.

48 **The area around the terminal where the planes park is called the apron.** In large airports, the passengers walk through a closed bridge directly onto the plane. When everyone is on board and settled in their seats, the doors close and the bridge pulls away. Then the plane moves slowly towards the runway along paths called taxiways. Sometimes a truck called a tug may tow a plane into position on the runway.

49 **The control tower ensures planes move safely around the airport.** Air traffic controllers keep track of all planes, both on the ground and when they take off and land. Their job is to make sure the planes are far enough apart to avoid accidents. The controller and pilots talk to each other by radio. The pilots wait for the controllers to give permission to take off or land.

Terminal building

I DON'T BELIEVE IT!

About 68 million passengers travel through London's Heathrow airport every year – that's over 180,000 every day. On average, planes take off and land at Heathrow at a rate of one every minute.

50 **Between flights, engineers check that everything is working correctly.** The plane is cleaned and loaded with fresh food and drink. Luggage is taken out of the hold and the luggage for the next flight is loaded. Tankers refill the onboard tanks with fuel. At regular intervals the planes are given more thorough checks to make sure they are always safe to fly.

Helicopters

Forward tilt

Backward tilt

Sideways tilt

Move up

Move down

Rotor hub

Rotor blade

Jet engine

Because they can move in any direction, helicopters are extremely useful. The pilot tilts the blades to change direction.

51 **A helicopter has whirling rotors to lift it into the air.** The rotors have blades that are shaped like long, narrow wings. The engine spins the blades. This lifts the helicopter in the same way as the wings on a plane. A tail rotor stops the helicopter spinning in the opposite direction to the main rotor.

52 **Helicopters can move in any direction – up, down, right, left, forwards or backwards.** This is done by changing the tilt of the rotor blades. To go forwards, the blades twist as they turn to give more lift at the back. This tilts the tail up and the helicopter moves forwards. It can also spin on the spot using the tail rotor. All this movement makes helicopters very difficult to fly.

53 Helicopters do not need a runway and can land anywhere as long as there is enough room. This makes them very useful, especially for rescuing people from the sea or mountains as they can hover in one place. While hovering, injured or stranded people can be lifted into the helicopter then carried to safety.

A helicopter can hover in one place to enable emergency workers to be lowered to accident scenes to help sick or injured people.

Tail rotor

Tailplane and fin

The Black Hawk military helicopter is used to carry troops and supplies during times of combat.

54 Some helicopters have double rotors to lift heavy loads. The Chinook helicopter has one rotor at the front and one at the back. One is slightly above the other and they turn in opposite directions. These helicopters can carry up to 55 soldiers or heavy military equipment slung underneath the helicopter.

Two large rotors give the Chinook helicopter extra lifting power to carry more people.

MAKE A WHIRLING ROTOR

Take a piece of paper 8 inches by 2 inches and fold it in half. Unfold it then cut from one short edge to the fold to make two strips. Fold these in opposite directions and put a paper clip on the other end. Watch it whirl like a helicopter rotor when you drop it.

Balloons and airships

55 **Balloons and airships can fly because they are lighter than air.** They are filled with a very light gas that tries to rise above the heavier air around it. If there is enough gas, it can lift the weight of the balloon or airship so that it floats up into the air.

The gas from a burner heats up the air inside a balloon, making the balloon float upwards.

Hot air balloons are flown for fun and are spectacular to watch. There are competitions held around the world where balloonists can compete in races.

LOOK FOR RISING HOT AIR

Watch the smoke from a BBQ or a bonfire. It always drifts upward. This is because the hot coal heats the air, which rises up above the surrounding colder air. The tiny smoke particles let us see the air rising.

56 **The air inside a hot air balloon is heated to make it lighter than the cool air around it.** Burners under the balloon heat the air inside, which spreads out and becomes lighter than the air outside. When the burners are turned on, the balloon rises. With the burners off, the balloon gradually falls as the air inside it cools. To fall more quickly, the pilot opens a vent (hole) and lets out some of the hot air. A balloon can't be steered, it goes where the wind blows it.

The first balloon to fly non-stop around the world was Breitling Orbiter 3.

58 **Weather balloons use a light gas called helium to go much higher than hot air balloons.** They measure temperatures and winds high above the clouds and send the information back by radio. Helium balloons have also broken records. In 1999 Bertrand Piccard and Brian Jones were the first people to fly a balloon around the world without stopping. It took them nearly 20 days in the Breitling Orbiter 3.

57 **Airships are also filled with light gas, but unlike balloons they have engines to steer them.** Modern airships use helium gas, which does not burn. The engines and propellers drive the airship and steer it with the help of rudders. There is a cabin for the pilot and passengers called a gondola. Airships are often seen hovering above events such as the Olympic Games, carrying TV cameras to give a view from the air.

This airship has been fitted with radar and tied to a ship. It will be used to spot icebergs beneath the water, and warn passing ships to steer clear of them.

Taking off vertically

The American V-z22 Osprey takes off like a helicopter, using its two rotors.

59 Vertical take-off planes do not need a long runway – they take off upward. There are two main types – planes that use propellers called tilt-rotor planes and those that have jet engines and are often called jump jets. These aircraft are mostly used as military planes. They can land on ships at sea and carry troops and equipment to army bases without a runway.

60 The V-22 Osprey is a plane with two large propellers at the ends of its wings. These lift the Osprey straight up into the air just like helicopter rotor blades. Then the engines swivel round at the ends of the wings so the propellers are tilted upright. The Osprey then flies like a normal plane. The propellers can also be tilted to allow the V-22 Osprey to hover like a helicopter.

A spinning fan inside the F-35B acts like a helicopter rotor.

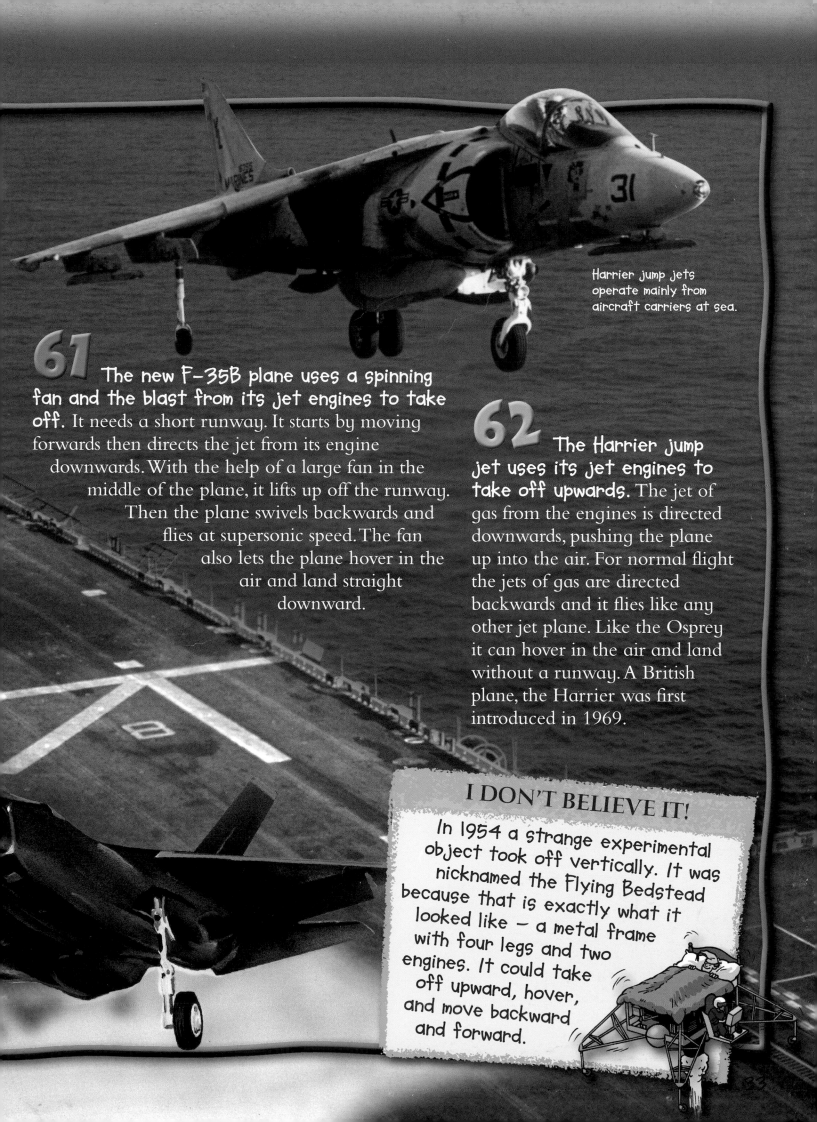

Harrier jump jets operate mainly from aircraft carriers at sea.

61 The new F-35B plane uses a spinning fan and the blast from its jet engines to take off. It needs a short runway. It starts by moving forwards then directs the jet from its engine downwards. With the help of a large fan in the middle of the plane, it lifts up off the runway. Then the plane swivels backwards and flies at supersonic speed. The fan also lets the plane hover in the air and land straight downward.

62 The Harrier jump jet uses its jet engines to take off upwards. The jet of gas from the engines is directed downwards, pushing the plane up into the air. For normal flight the jets of gas are directed backwards and it flies like any other jet plane. Like the Osprey it can hover in the air and land without a runway. A British plane, the Harrier was first introduced in 1969.

I DON'T BELIEVE IT!

In 1954 a strange experimental object took off vertically. It was nicknamed the Flying Bedstead because that is exactly what it looked like — a metal frame with four legs and two engines. It could take off upward, hover, and move backward and forward.

Planes for war

63 In war, planes are used to attack the enemy, drop bombs, and carry troops and equipment. Modern military planes often act as both fighter and bomber. They fly fast enough to attack and escape from danger while carrying bombs and missiles. Large bombers can usually fly longer distances than fighter planes.

After dropping its bombs a fighter plane speeds away. Fighters are usually small and agile.

64 Military planes have different weapons depending on the mission. They carry rockets, missiles and bombs as well as guns. Many missiles and bombs can find their own way to the target. Missiles have a rocket engine to home in on the target. "Smart" bombs have no engine. Instead they glide down using fins to steer onto the target. Pilots can fire their weapons at more than one target at the same time.

A fighter plane (far right) flies next to a bigger tanker plane (below) for refuelling.

65 Fighter planes are very difficult to fly. Pilots learn to fly them in simulators before they fly the real plane. Simulators are machines that use computers to make the pilot feel as if he is flying a real plane.

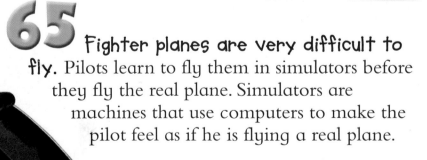

Specialized helmets allow fighter pilots to communicate with other pilots and air controllers. The helmet also supplies the pilot with air to breathe.

66 **The armed forces use huge cargo planes to carry troops and heavy equipment.** They deliver tanks and all the other supplies and equipment needed by an army wherever it is fighting. Cargo planes need a runway to land on, so helicopters take over from planes to carry men and equipment around the battlefield.

Global Hawk spy plane is a robot that operates without a pilot.

67 **Planes can spy on the enemy.** From the air, spy planes use cameras and spying equipment to give a picture of what is happening on the ground and where enemy planes are. A control center on the plane keeps its own forces informed and tells them what and where to attack. Spy planes also fly over the ocean to spot submarines and warn naval ships.

68 **Many military planes can refuel while flying.** They often have to fly long distances and there may not be anywhere they can land to refuel. Large tanker planes carrying fuel fly alongside. A pipe links the two planes and delivers fuel from the tanker. This allows the fighter or bomber to fly much further.

I DON'T BELIEVE IT!

Most military planes have ejection seats so the pilot can escape if the aircraft is hit. The whole seat with the pilot sitting in it is catapulted out of the cockpit. A rocket shoots it upward, then a parachute opens to lower it gently to the ground.

Working planes

69 **Transport planes carry cargo around the world.** Inside there are no seats, just a huge area to be filled with containers of goods. A Boeing 747 jumbo jet can hold as much as five large trucks. It is loaded through hatches on the side.

The whole nose of this C-5 Galaxy cargo plane opens up to load very large objects.

A Russian Antonov An-225, the biggest plane in the world. It is used to carry extremely large loads. Here it is transporting a Russian Space Shuttle.

70 **The biggest plane is the Russian Antonov An-225.** It can carry a cargo of 250 tons and was designed to transport the Russian Space Shuttle. The Antonov has six engines and its wings measure almost 300 feet from tip-to-tip. Two Antonovs have been built, but only one has ever flown.

A firefighting plane dumps its load of red fire retardant in an attempt to put out a forest fire. This is a special substance that will stop the fire spreading.

73 Photographs taken from a flying plane give a good view of the ground. These photos can help people to draw maps. They also help historians by finding forgotten villages and roads. Slight bumps left by the burial of old ruins show up more clearly from the air than from the ground.

71 Planes and helicopters can help put out forest fires by dropping water on the fire. Special water-bombing planes fly very low over the sea or a lake and scoop up water. They then fly over the fire and drop the water. Helicopters carry a large bucket underneath and use this to scoop up and drop the water.

72 Helicopters and small planes can act as ambulances. If a person is badly injured in a road accident, an air ambulance helicopter may be called. The patient is put on a stretcher and whisked away to a hospital in the helicopter. In Australia, doctors use small planes to visit remote farms and villages that it would take too long to reach by car.

74 The police use helicopters to watch the traffic and chase criminals. Police in a helicopter can spot an escaping criminal and guide the police on the ground to help catch him, particularly in a car chase. The helicopters also fly over busy roads, reporting back on accidents and traffic jams.

Planes for fun

75 **Some people fly planes just for fun or sport.** They use small planes or gliders that take off from small airfields. They do not need a long concrete runway – a strip of mown grass is often good enough. Most small planes can only carry a few passengers and some only have a seat for the pilot.

The UK Air Force Red Arrows aerobatic team has been giving displays since 1965.

Gliders are made from light materials. They can stay in the air for hours if conditions are right.

77 **The smallest kind of plane is called a microlight.** This plane is very light, with just a wing, a propeller engine and one or two seats in an open cockpit. Some look like very small planes but others look more like large kites, with wings that fold up. Pilots have to pass a test to fly a microlight just like they would for any other sort of plane.

76 **Gliders are planes with no engines.** The shape of a glider is long and thin to slice through the air, with long wings for lift. To become airborne, they are towed along the ground until they are moving fast enough to take off. Some are towed into the air by a plane, which takes off pulling the glider behind it. In the air, the plane drops the tow rope and the glider flies on its own gradually dropping to the ground.

A small propeller behind the pilot drives this microlight.

78 Paragliders and hang gliders do not have engines, and the pilots take off from the top of a hill or cliff. A hang glider is like a large kite with a person hanging below, strapped into a harness. A paraglider is more like a parachute, with a canopy that holds air to make a wing shape. Pilots take off by running into the wind, and gradually glide down to the ground.

79 Gliders use rising currents of air to fly higher. They fly near hills where the wind flows up and over the hill. It lifts the glider up higher so it can stay in the air longer. Gliders also ride on currents of warm air called thermals. These form when air is heated by the warm ground and rises like the hot air in a balloon.

80 Flying a small plane in complicated loops, rolls and turns is called aerobatics. Pilots often perform aerobatics for competitions. Groups of planes flying in formation with coloured smoke streaming out behind them put on spectacular displays at air shows. The pilots need lots of practice and skill to do this safely. Helicopters can also perform aerobatic displays.

MAKE A GLIDER

Fold a piece of paper in half lengthwise then open it out flat. Fold the top corners down to the center, making an arrow shape. Then fold the two sides in again. Fold the paper back along the original center fold. Hold the glider's nose with the wings open flat. Throw it to see how far it flies. Try again with a paper clip on the nose.

Planes at sea

81 Military aircraft go to sea on board huge ships called aircraft carriers. These can take planes as close as possible to war zones. The planes use the aircraft carrier like an airport, taking off for a mission then returning to land and refuel. Many planes can fold their wings up when not flying so that more can fit onboard.

82 The top deck of an aircraft carrier is the flight deck where the planes take off and land. It is like a runway but not as long, so a catapult shoots the planes forward, giving them extra speed to fly from the deck. After landing, the plane is brought to a standstill by a wire hooked across the deck. Below the flight deck is a hangar deck where planes are stored. They go up and down between the decks in a huge lift.

83 Vertical take-off planes can operate from smaller aircraft carriers. They do not need a runway just enough space on the deck to take off and land again. These smaller carriers often have a ramp at the end of the deck for planes that need a short runway to help them take off.

TRUE OR FALSE?

1. Seaplanes can land on land.

2. On an aircraft carrier, planes have wings that fold up.

3. A helipad is an area for planes and helicopters to land and take off.

Answers:
1. False 2. True 3. True

Planes park on the flight deck of an aircraft carrier between flights.

A seaplane floats on the water ready for take-off. These small planes are sometimes used by coast guards in rescue operations.

85 **Seaplanes can land on water.** These small planes have floats instead of wheels. The floats rest on the surface so the plane is out of the water. Seaplanes can only land and take off if the water is calm. They are sometimes used for flying between islands, or in remote areas where there are lakes to land on but few runways.

84 **Helicopters are also used on ships.** They only need a small platform called a helipad for taking off and landing. They cannot travel as far or as fast as a plane, but they can ferry people and equipment and act as look outs. Oil rigs out at sea often depend on helicopters to bring new crews and supplies from the mainland.

Helipads provide a landing place at sea.

Flying faster than sound

86 **Supersonic planes fly faster than sound.** Sound travels extremely fast – when you clap your hands the sound moves out in all directions and people hear it almost immediately. Supersonic planes travel faster than this. When a plane starts to travel faster than sound, we say it is breaking the sound barrier.

An American F-16 Fighting Falcon can fly twice as fast as sound.

88 **Many military fighter planes are built to fly faster than sound.** Their engines often have afterburners for more power and speed. These burn extra fuel in the stream of hot gas coming out of the engine. This gives a plane more power for take-off, or for a short burst of extra speed, and allows it to fly for longer at supersonic speeds.

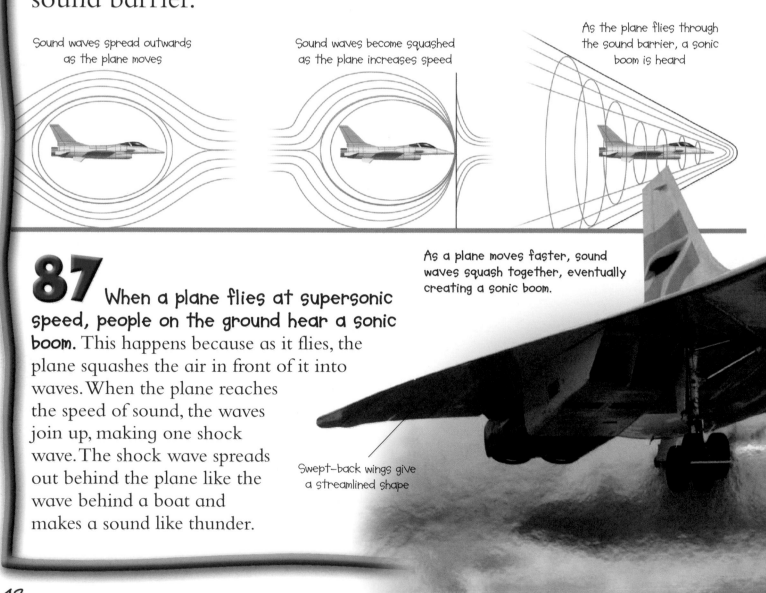

Sound waves spread outwards as the plane moves

Sound waves become squashed as the plane increases speed

As the plane flies through the sound barrier, a sonic boom is heard

87 **When a plane flies at supersonic speed, people on the ground hear a sonic boom.** This happens because as it flies, the plane squashes the air in front of it into waves. When the plane reaches the speed of sound, the waves join up, making one shock wave. The shock wave spreads out behind the plane like the wave behind a boat and makes a sound like thunder.

As a plane moves faster, sound waves squash together, eventually creating a sonic boom.

Swept-back wings give a streamlined shape

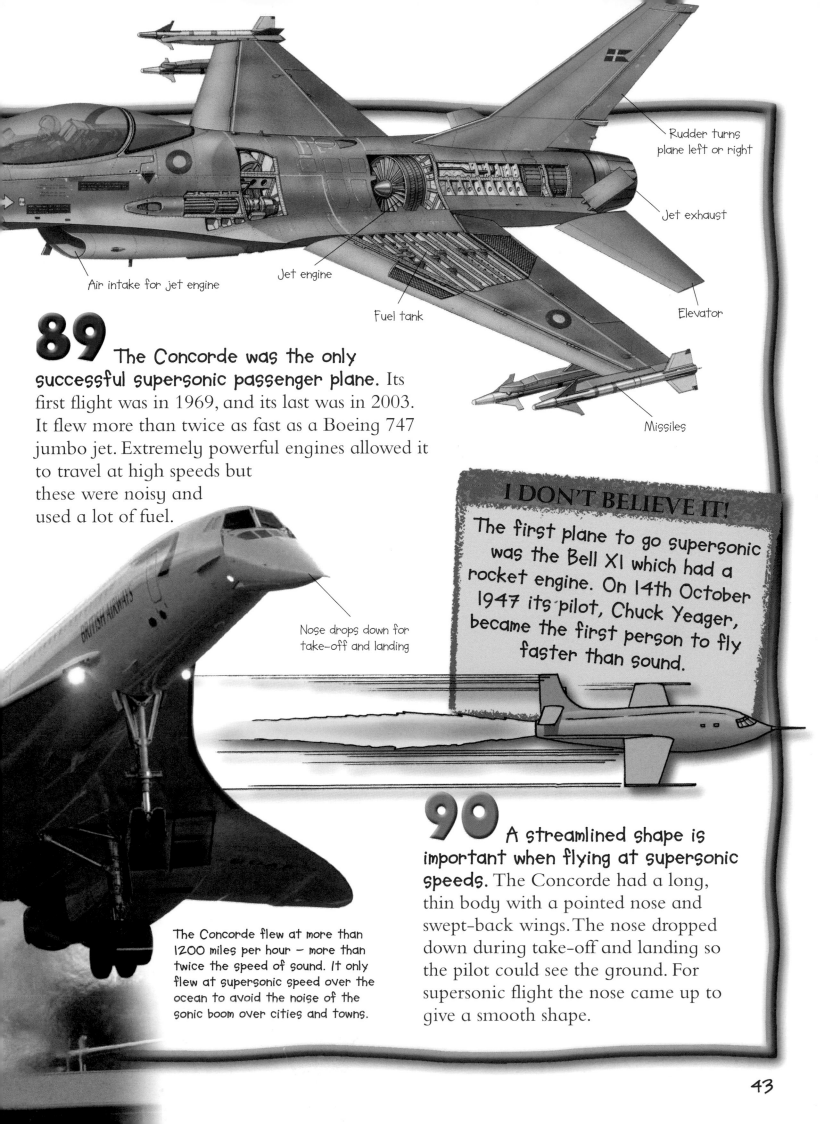

Rudder turns plane left or right

Jet exhaust

Elevator

Air intake for jet engine

Jet engine

Fuel tank

Missiles

89 The Concorde was the only successful supersonic passenger plane. Its first flight was in 1969, and its last was in 2003. It flew more than twice as fast as a Boeing 747 jumbo jet. Extremely powerful engines allowed it to travel at high speeds but these were noisy and used a lot of fuel.

Nose drops down for take-off and landing

I DON'T BELIEVE IT!

The first plane to go supersonic was the Bell XI which had a rocket engine. On 14th October 1947 its pilot, Chuck Yeager, became the first person to fly faster than sound.

The Concorde flew at more than 1200 miles per hour – more than twice the speed of sound. It only flew at supersonic speed over the ocean to avoid the noise of the sonic boom over cities and towns.

90 A streamlined shape is important when flying at supersonic speeds. The Concorde had a long, thin body with a pointed nose and swept-back wings. The nose dropped down during take-off and landing so the pilot could see the ground. For supersonic flight the nose came up to give a smooth shape.

Flying in space

91 **Powerful rockets can fly into space.** There is no air in space, so wings are of no use. Rockets are pushed upwards by their powerful engines. Also, without air there is no drag to slow the rocket. Once started by a boost from its motor, the rocket keeps going, only needing extra boosts to change direction or speed.

2. Booster rockets fall away

1. Rocket engine's fire

The Shuttle's rocket engines and boosters give it enough speed to reach space.

The Ariane 5 rocket launches satellites into space. A new rocket is built for each launch.

Liquid oxygen tank

Liquid hydrogen tank

92 **Rocket engines are similar to jet engines.** Inside the engine the fuel burns, making hot gases rush out through a nozzle at the back, and the rocket shoots forwards. However a rocket carries its own oxygen gas to burn the fuel. A jet engine uses oxygen from the air. This means a jet engine only works in air, but a rocket engine works in air and space.

The solid fuel in the Shuttle's booster rockets burns very rapidly for maximum thrust.

3. Fuel tank separates

93
The Space Shuttle takes off as a rocket. It has three rocket engines, but it also uses two huge booster rockets. These only fire for two minutes before dropping back to Earth. The engines use fuel from a separate tank that also falls away when the Shuttle reaches space. The booster rockets land in the sea and are used again, but the tank burns up as it drops into the atmosphere.

4. Return to Earth

5. Touch down

94
The Shuttle lands on a runway like a huge glider. When the Shuttle returns to Earth it does not use its rocket engines. It swoops gently downwards and uses its wings and tail to slow and guide it towards its long runway. It touches down much faster than an airliner and uses a parachute to help it slow down and stop.

Exhaust nozzle

Orion does not need wings like the Shuttle because it will not land on a runway.

Solid fuel booster

95
A new spacecraft called Orion will replace the Shuttle. It will be launched by a rocket called Ares, which is similar to the Shuttle's boosters. Four to six astronauts will travel in the cone-shaped capsule. They will be able to go to the International Space Station, or the Moon – or even the planet Mars. When they return to Earth they will float down using parachutes and land in the sea.

45

Strange planes

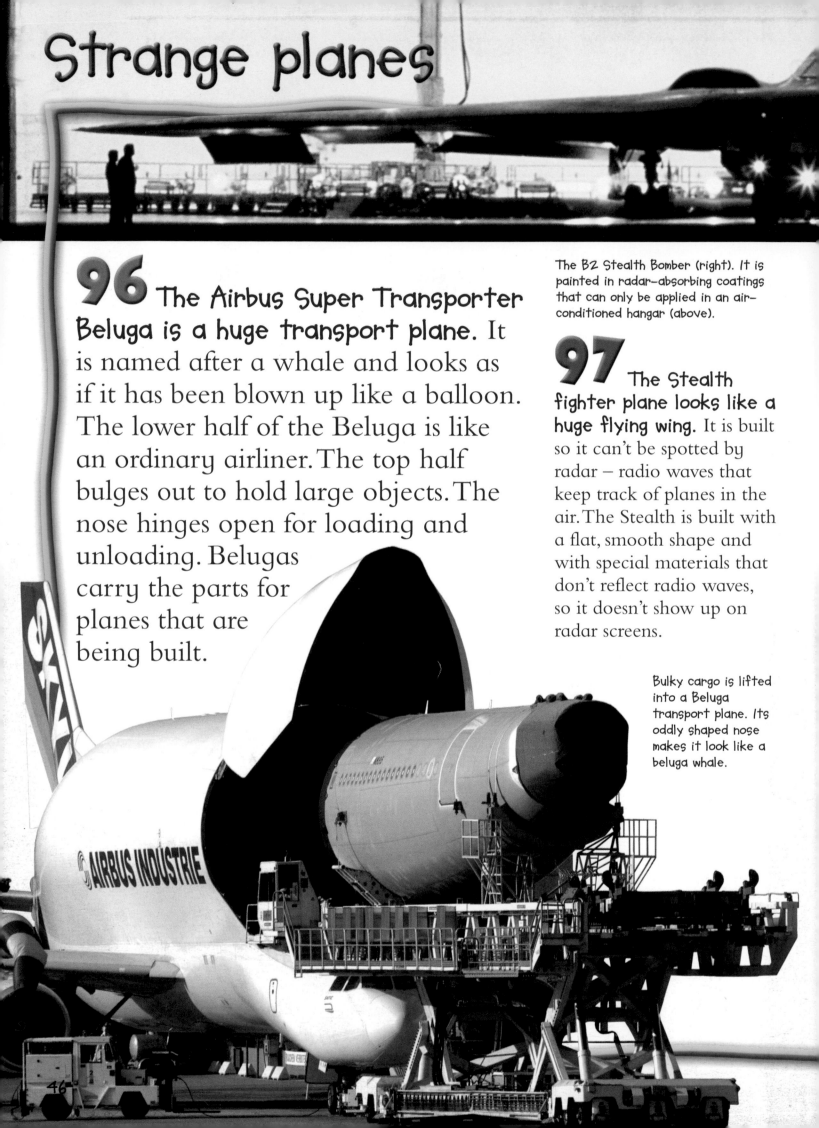

96 The Airbus Super Transporter Beluga is a huge transport plane. It is named after a whale and looks as if it has been blown up like a balloon. The lower half of the Beluga is like an ordinary airliner. The top half bulges out to hold large objects. The nose hinges open for loading and unloading. Belugas carry the parts for planes that are being built.

The B2 Stealth Bomber (right). It is painted in radar-absorbing coatings that can only be applied in an air-conditioned hangar (above).

97 The Stealth fighter plane looks like a huge flying wing. It is built so it can't be spotted by radar – radio waves that keep track of planes in the air. The Stealth is built with a flat, smooth shape and with special materials that don't reflect radio waves, so it doesn't show up on radar screens.

Bulky cargo is lifted into a Beluga transport plane. Its oddly shaped nose makes it look like a beluga whale.

98 **The Sun can provide enough power to fly a plane.** The Solar Challenger plane had long wings covered in more than 16,000 solar cells. These used sunlight to make electricity and drive a propeller. It flew from France to England in 1981. *Helios* was a strange-looking plane – just a huge wing with 14 propellers along the front. These ran on electricity from solar cells covering the wings.

99 **The Gossamer *Albatross* flew using human power.** It was a very light plane with long, thin wings – just a flimsy frame wrapped in plastic. The pilot sat inside a tiny cabin pedalling (like a bicycle) to turn a propeller that pushed the plane forwards. It flew across the English Channel from England to France in 1979.

Helios was a simple wing covered in solar panels, which powered its 14 propellers. It reached a top speed of 20 miles per hour and was made by NASA.

I DON'T BELIEVE IT!

The Gossamer *Albatross* weighed less than its pilot. Together they weighed 220 pounds but only 70 pounds of that was the actual plane. It flew only 5 feet above the ground.

100 *Predator* is a robot spy plane. It is the same size as a small plane but with no cabin or pilot. It flies by remote control like a model plane. Cameras show its controllers a view of the enemy. It can also carry missiles to attack enemy targets.

Index

Entries in **bold** refer to main subject entries. Entries in *italics* refer to illustrations.